Review of EPA Homeland Security Efforts

Safe Buildings Program Research Implementation Plan

Committee on Safe Buildings Program

Board on Chemical Sciences and Technology

Division on Earth and Life Studies

NATIONAL RESEARCH COUNCIL
OF THE NATIONAL ACADEMIES

THE NATIONAL ACADEMIES PRESS
Washington, D.C.
www.nap.edu

THE NATIONAL ACADEMIES PRESS 500 Fifth Street, N.W. Washington, DC 20001

NOTICE: The project that is the subject of this report was approved by the Governing Board of the National Research Council, whose members are drawn from the councils of the National Academy of Sciences, the National Academy of Engineering, and the Institute of Medicine. The members of the committee responsible for the report were chosen for their special competences and with regard for appropriate balance.

Support for this study was provided by the U.S. Environmental Protection Agency under contract number 68-C-03-037.

All opinions, findings, conclusions, or recommendations expressed in this publication are those of the author(s) and do not necessarily reflect the views of the organizations or agencies that provided support for this project.

International Standard Book Number 0-309-09104-7 (Book)
International Standard Book Number 0-309-52823-2 (PDF)

Additional copies of this report are available from:

National Academy Press
500 Fifth Street, N.W.
Box 285
Washington, DC 20055
800-624-6242
202-334-3313 (in the Washington metropolitan area)
http://www.nap.edu

Printed in the United States of America

THE NATIONAL ACADEMIES

Advisers to the Nation on Science, Engineering, and Medicine

The **National Academy of Sciences** is a private, nonprofit, self-perpetuating society of distinguished scholars engaged in scientific and engineering research, dedicated to the furtherance of science and technology and to their use for the general welfare. Upon the authority of the charter granted to it by the Congress in 1863, the Academy has a mandate that requires it to advise the federal government on scientific and technical matters. Dr. Bruce M. Alberts is president of the National Academy of Sciences.

The **National Academy of Engineering** was established in 1964, under the charter of the National Academy of Sciences, as a parallel organization of outstanding engineers. It is autonomous in its administration and in the selection of its members, sharing with the National Academy of Sciences the responsibility for advising the federal government. The National Academy of Engineering also sponsors engineering programs aimed at meeting national needs, encourages education and research, and recognizes the superior achievements of engineers. Dr. Wm. A. Wulf is president of the National Academy of Engineering.

The **Institute of Medicine** was established in 1970 by the National Academy of Sciences to secure the services of eminent members of appropriate professions in the examination of policy matters pertaining to the health of the public. The Institute acts under the responsibility given to the National Academy of Sciences by its congressional charter to be an adviser to the federal government and, upon its own initiative, to identify issues of medical care, research, and education. Dr. Harvey V. Fineberg is president of the Institute of Medicine.

The **National Research Council** was organized by the National Academy of Sciences in 1916 to associate the broad community of science and technology with the Academy's purposes of furthering knowledge and advising the federal government. Functioning in accordance with general policies determined by the Academy, the Council has become the principal operating agency of both the National Academy of Sciences and the National Academy of Engineering in providing services to the government, the public, and the scientific and engineering communities. The Council is administered jointly by both Academies and the Institute of Medicine. Dr. Bruce M. Alberts and Dr. Wm. A. Wulf are chair and vice chair, respectively, of the National Research Council.

www.national-academies.org

COMMITTEE ON SAFE BUILDINGS PROGRAM

STEVEN R. TANNENBAUM, Massachusetts Institute of Technology, Chair
WILLIAM P. BAHNFLETH, The Pennsylvania State University
R. JOHN COLLIER, Harvard Medical School
DAVID A. EDWARDS, Harvard University
LEON P. GLICKSMAN, Massachusetts Institute of Technology
LYNNE HABER, Toxicology Excellence for Risk Assessment
SANGTAE KIM, Eli Lilly and Company
CHARLES E. KOLB, JR., Aerodyne Research, Inc.
ELLEN RABER, Lawrence Livermore National Laboratory
WILLIAM S. REES, JR., Georgia Institute of Technology
RICHARD G. SEXTRO, Lawrence Berkeley National Laboratory
KENT J. VOORHEES, Colorado School of Mines

Staff

ANDRIA L. HOBBS, Christine Mirzayan Intern
ERIC L. SHIPP, Postdoctoral Fellow
SYBIL A. PAIGE, Administrative Associate
DAVID C. RASMUSSEN, Program Assistant
DOROTHY ZOLANDZ, Director, Board on Chemical Sciences and Technology

BOARD ON CHEMICAL SCIENCES AND TECHNOLOGY

Staff

Acknowledgment of Reviewers

This report has been reviewed in draft form by individuals chosen for their diverse perspectives and technical expertise, in accordance with procedures approved by the National Research Council's Report Review Committee. The purpose of this independent review is to provide candid and critical comments that will assist the institution in making the published report as sound as possible and to ensure that the report meets institutional standards for objectivity, evidence, and responsiveness to the study charge. The review comments and draft manuscript remain confidential to protect the integrity of the deliberative process. We wish to thank the following individuals for their review of this report:

W. Emmett Barkley, Howard Hughes Medical Institute
Joan B. Berkowitz, Farkas Berkowitz & Company
Frank P. Crimi, Lockheed Martin Advanced Environmental Systems Company (retired)
Lynn M. Hildemann, Stanford University
David E. Neff, Colorado State University
George W. Parshall, DuPont Company (retired)
Eugene Sevin, Consultant, Lyndhurst, Ohio
Michael A. Wartell, Indiana University—Purdue University Fort Wayne

Although the reviewers listed above have provided many constructive comments and suggestions, they were not asked to endorse the conclusions or recommendations nor did they see the final draft of the report before its release. The review of this report was overseen by Robert Frosch, Harvard University. Appointed by the National Research Council, he was responsible for making certain that an independent examination of this report was carried out in accordance with institutional procedures and that all review comments were carefully considered. Responsibility for the final content of this report rests entirely with the authoring committee and the institution.

Contents

Executive Summary

The U.S. Environmental Protection Agency (EPA) has important and challenging tasks in the areas of building protection and decontamination. The EPA initiated its Safe Buildings Program as a comprehensive response to a chemical or biological attack on a civilian or public sector facility. The Safe Buildings Research Implementation Plan (RIP) was devised to guide EPA research and demonstrates efforts in this area over the course of three years—ending in fiscal year 2005.

The Research Implementation Plan has four research foci: detection, containment, decontamination, and disposal.

- The detection research focuses on research areas that deal with real-time detection and "detect-to-treat" types of detection of biological and chemical agents in the event of an attack.
- The containment research focus is on the development and testing of methods to prevent the spread of contaminants within buildings in order to protect building occupants, first responders, and decontamination crews. The overall objective is to reduce or eliminate the impact of a chemical or biological attack on building occupants as well as to provide techniques and guidance to determine the efficacy of chemical and biological protection measures for new and existing buildings.
- The decontamination research area focuses on providing the tools, techniques, technologies, and guidance needed to decontaminate a building subsequent to a chemical or biological attack.
- The final research area, disposal, aims to provide guidance for disposal of materials contaminated by chemical and biological agents or materials that have been contaminated as a result of the decontamination efforts.

At the request of the EPA, the National Research Council formed a committee to provide a review of the Research Implementation Plan of the National Homeland Security Research Center's Safe Buildings Program. The committee met twice (May 13-14, 2003, and July 10, 2003) to learn about the program's context, goals, and content through presentations and discussions and to review the Research Implementation Plan and associated background materials. Committee-only sessions were held at both meetings to begin the review task and to provide an opportunity for committee members to discuss and refine the review comments and concerns.

As it began addressing the Statement of Task (Appendix A), the committee found that some of the questions were not really appropriate due to the relative newness of the Safe Buildings Program and the RIP. Specifically, the committee found that overarching issues were the three-year time frame and the amount of work EPA has proposed to accomplish. Rather than perform a detailed review of every proposal in the RIP, the committee felt it was more important to provide EPA with key areas that it can impact during the remaining time in the program by establishing a prioritization of the four major program areas. For similar reasons it was not useful to discuss sequencing of the projects, as a three-year time frame does not allow for programs beyond the short-term to be addressed by EPA. However, recommendations for establishing a longer-term project are mentioned in the report. Given the committee's determinations in regards to handling the questions in the Statement of Task, it seemed unnecessary to comment on the presentation and structure of the RIP.

The overarching findings and recommendations of the committee are presented below.

EPA has correctly identified the major research areas essential for the Safe Buildings Program.

The primary areas of research associated with an effective building decontamination strategy are presented in the Safe Buildings Program Research Implementation Plan: detection, containment, decontamination, and disposal.

The duration of the current program is insufficient to deal with all the tasks and goals presented in the Research Implementation Plan.

As specifically noted in the various findings and recommendations, the program time frame (scheduled to end at the end of fiscal year 2005) is too short to effectively accomplish all the goals set forth in the Research Implementation Plan. The proposed plan covers an extensive area of new research, of which the scope and breadth is too large to accomplish in the allotted time frame. Given current resources and the extramural collaborations of EPA, it is unrealistic to expect results in all areas of proposed research in the remaining time.

The current effort should include a planning function for a potential longer-term research program to address unmet needs in technical areas.

The short time frame proposed in the Research Implementation Plan is an overarching concern; accordingly, the committee has made specific recommendations for tasks that can reasonably be completed in three years or less. One of these tasks should be the development of a coordinated program for a long-term research and development effort focusing on the agency's strengths. In some areas, the committee has provided guidance or recommendations for long-term research activities.

In the short term, the program should focus almost exclusively on decontamination and disposal issues and other parts of the program should be subordinate to decontamination and disposal.

EPA has expertise in the area of decontamination and disposal. The agency should focus the remaining time toward improving these aspects of its work. Activities in the remaining areas—detection and containment—should technically support decontamination and disposal in a logical manner to achieve results within the prescribed time period of the program. The committee has made specific recommendations for these areas.

1

Introduction

In light of contamination of the Hart Senate Office Building, the Brentwood postal facility in Washington, D.C., and a few other buildings with *Bacillus anthracis* (anthrax) in the fall of 2001, Congress has allocated \$50 million to the United States Environmental Protection Agency (EPA) for research on issues relating to the safety of building occupants when a chemical or biological attack occurs. In response EPA has initiated a comprehensive research program focusing on building security, building protection, and the safe and efficient decontamination of buildings when hazardous materials are deliberately introduced. Over the remaining time of the program (to the end of fiscal year 2005) EPA intends to provide research and technical assistance on methods, technologies, equipment, and tools required to protect built structures from deliberate attacks with biological warfare agents, chemical warfare agents, and toxic industrial chemicals through the National Homeland Security Research Center's Safe Buildings Program.

EPA is targeting the products of this program for use by building owners and managers, architects, emergency responders, decontamination crews, state and local public health officials, EPA program and regional offices, and other agencies involved in protecting human health and the environment. The Safe Building Program addresses three issues.

1. How can building occupants be protected during a terrorist attack that contaminates the indoor air?
2. How can contaminated buildings be safely, efficiently, and cost-effectively restored?
3. How can decontamination information be effectively conveyed to building owners and managers, emergency responders, decontamination crews, and federal regulatory groups?

The cornerstone of EPA's Safe Buildings Program is the Research Implementation Plan (RIP), a document that outlines the various research areas, proposals, and research timeframes for the program. It is designed to address all areas of building protection and decontamination needs, identify highest priority needs for immediate action, and determine how best to meet these needs.

The plan comprises four research areas that are considered separately.

1. Detection research seeks to ensure that effective sampling and analysis tools are available to conduct a thorough building decontamination. It will also develop monitoring techniques that can be used to detect an attack and provide safe re-occupancy of a building.
2. Containment research aims to develop methods

to prevent the spread of contaminants in order to protect building occupants, first responders, and decontamination crews as well as provide techniques and guidance to determine the efficacy of chemical and biological protection measures for new and existing buildings.

3. Decontamination research will provide the tools, techniques, technologies, and guidance needed to decontaminate a building subsequent to a chemical or biological attack.
4. Disposal research will provide guidance for disposal of materials contaminated by chemical and biological agents or materials that have been contaminated as a result of decontamination.

EPA commissioned the National Academies to assemble a committee to address the following questions with regards to the Safe Buildings Program Research Implementation Plan:

- From the expert committee's viewpoint, has the Research Implementation Plan completely and accurately identified important issues and needs in the buildings security arena? If the answer is no, what issues and needs should be added?
- From the expert committee's viewpoint, are the needs accurately prioritized and sequenced within the issues? If the answer is no, what would be the recommended adjustments and why is this the case?
- From the expert committee's viewpoint, are the projects recommended for funding over the next three years under each need appropriately sequenced to move to the ultimate product or products identified in the Appendix? If the answer is no, what would the committee recommend as adjustments and why is this the case?
- Overall, from the expert committee's viewpoint, what changes to the RIP would be recommended to improve its presentation in terms of content and structure so as to more clearly convey the buildings security research and technical support program that is described?

EPA gave the committee a draft of EPA's Safe Buildings Program Research Implementation Plan,[1] which consisted of a program overview along with work proposals in the four outlined areas (detection, containment, decontamination, and disposal). In addition, the plan stressed the completion of EPA research projects within the allotted three years. After a presentation by the project leaders in each of the four areas, the committee concluded it needed the following additional information from EPA in order to perform a fair and accurate assessment of the RIP:

- Detailed information on the threat scenario analysis and its relation to the major research programs, including a matrix describing the relevance of research proposals implemented as a result.
- More information about the gap analysis performed in order to ensure minimal overlap between research areas already in place.
- A list of the research proposals already underway.
- A document detailing the lessons learned from the recent anthrax decontamination efforts

The content of this report is the committee's review of EPA's RIP. The critiques and ideas expressed in this document are intended to aid EPA in determining the most efficient strategy to implement its research plan in a manner that provides the most success to EPA and to the nation. Chapter 2 of this report contains an assessment of each RIP area, identifying both strengths and weaknesses. The committee's findings and recommendations are outlined in Chapter 3.

[1]EPA, 2003, Research Implementation Plan—Safe Buildings Program (Draft), presented to the committee on May 13, 2003, by N. Adams, Environmental Protection Agency, Research Triangle Park, NC.

2

Assessment

NRC ASSESSMENT CONTEXT AND CRITERIA

The committee recognizes that the building protection and decontamination tasks the Environmental Protection Agency (EPA) has been given are both exceedingly important and extremely challenging. Analyses of the decision processes required to effectively handle a chemical or biological attack on a civilian or public sector facility demonstrate that a wide range of time-critical information and technical capabilities are required to identify, respond to, characterize, and decontaminate or remediate the resulting damage (Raber et al. 2002). Evaluations of the responses to recent anthrax contamination of congressional and Post Office facilities clearly demonstrate the severe management and technical difficulties posed by even a modest biological attack (EPA, 2002; GAO, 2003b). While some straightforward measures have been identified to protect buildings from chemical or biological attacks (CDC/NIOSH, 2002), more comprehensive attack warning, evaluation, and containment strategies will require extensive advances in building systems and their components.

The EPA Safe Buildings Program was conceived as a comprehensive response to these challenges.[1] The program is motivated by a threat analysis performed by the agency with input from a range of relevant federal sources.[2] However, the current program is constrained by the time frame (three years total, approximately 30 months remaining at the start of this review process). Given the time currently available to the program, the committee has attempted to identify critical safe building research and development challenges that might be substantially completed within the time and cost constraints as well as to note those that will clearly require longer time frames or more substantial budgets or both to ensure significant progress.

It is also apparent to the committee that execution of any safe buildings program will require an extremely wide range of chemical, biological, engineering, and social science expertise. EPA's traditional role of protecting human health and ecosystem viability from insult by industrial and commercial products and byproducts has resulted in many staff and contractor capabilities that are very pertinent to the safe buildings challenge. However, the agency and its contractors may not have all the skills required to specify, develop, evaluate, and demonstrate desirable safe buildings

[1]EPA, 2003, Research Implementation Plan—Safe Buildings Program (Draft), presented to the committee on May 13, 2003, by N. Adams, Environmental Protection Agency, Research Triangle Park, NC.

[2]EPA, 2003, Analysis of Potential Scenarios for Terrorist Attacks on Buildings with Biological or Chemical Agents (Draft), Lemieux, P., Adams, N., and Sparks, L., presented to the committee on July 10, 2003, by N. Adams, Environmental Protection Agency, Research Triangle Park, NC.

technology components and systems. The committee has attempted to identify Safe Buildings Program tasks that require technical expertise in areas where the agency has strong and in some cases even unique capabilities; as well as to flag those tasks for which the agency will need to develop new skills or acquire new collaborators to succeed.

Given the time scale for the current Safe Buildings Program, the committee recommends that priority be given to those tasks that can reasonably be completed within the program time frame and that draw most heavily on the technical expertise of EPA's laboratories and current contractors. The committee also recommends that longer-range tasks and those that require the development of expanded capabilities be deferred in the short term or undertaken only in close collaboration with agencies and organizations with the required capabilities.

The committee notes that the threat scenario used to define the Safe Buildings Program[3] should be used to identify priority needs and to set program goals. The Safe Buildings Program will likely be most successful when specific, realistic program goals are set early in the program and when success or failure is consistently and systematically measured against those goals as the program evolves. In the assessments of each Safe Buildings Program element presented below the committee identifies goals and measures of success.

EPA also has a list for measures of success within the three-year time frame[4]:

- identify the most important threats (key drivers of the program);
- develop and define appropriate and effective methods for detecting, containing, and decontaminating events relating to these top threats;
- produce methods and guidance for understanding and predicting human health risks;
- lead the development of clean-up standards; and
- distribute methods/guidance to building owners, responders, and appropriate public officials (An additional measure of success will be successful use of these technologies and the guidance by our customers.)

The committee's assessment evaluates the program in the context of these success factors as well as others identified by the committee in its discussions.

DETECTION

Different Approaches to Chemical and Biological Detection and Characterization Systems

In 1999 the National Research Council (1999a) discussed numerous biological and chemical detection systems, many of which have since been improved. The advantages and disadvantages of these systems should be incorportated into EPA's strategy for characterization, decontamination, and recovery. The technologies most applicable to monitoring and notification for any agent must be sensitive enough to detect agent concentrations at or below health risk levels, specific enough to provide acceptable false-alarm rates, and prompt enough for notification consistent with effective medical response. This generally equates to a critical response time of minutes for chemical agents and toxic industrial chemicals and days for most biological agents. For attacks using chemical warfare agents, symptoms are typically prompt, with coughing, choking, distress, and sometimes death occurring as soon as seconds after exposure. Detectors must respond in nearly real time to minimize exposures and guide medical intervention. Following a chemical attack it is likely that well-established hazardous material operations would be conducted during the notification phase. Chemical sensors that respond quickly are already available commercially from several sources, and analytic instruments incorporating preconcentration methods are already available to detect and identify chemical warfare agents at or below known health risk limits. Many of these methods have been developed to support either the military or the current international chemical weapons treaty verification program. After the Aum Shinrikyo attack in Tokyo, strategies have been investigated for detecting and responding to chemical warfare agents in semi-enclosed structures, such as subways and airports. In the United States,

[3]EPA, 2003, Analysis of Potential Scenarios for Terrorist Attacks on Buildings with Biological or Chemical Agents (Draft), Lemieux, P., Adams, N., and Sparks, L., presented to the committee on July 10, 2003, by N. Adams, Environmental Protection Agency, Research Triangle Park, NC.

[4]EPA, 2003, Research Planning in the Safe Buildings Program—Report to the National Academies, presented to the committee on June 30, 2003, by N. Adams, Environmental Protection Agency, Research Triangle Park, NC, p. 29.

much of this work has been conducted through agencies that are now part of the Department of Homeland Security. The NRC Committee on Safe Buildings Program recommends that EPA continue to utilize these ongoing studies to aid implementation of the Safe Buildings Program.

For biological attacks detection can be based on environmental monitoring, epidemiological monitoring, or an unusual diagnosis, as was the case with the inhalation anthrax incidents. Actual detection of an event is much more difficult and the time delay is significant when compared with chemical incidents. Several projects both within EPA and other agencies are underway to enhance early detection and notification, and some environmental monitoring systems have already been deployed in major U.S. metropolitan areas (Cole, 2003). Because many biological threat agents are zoonotic (outbreaks naturally occurring in animals), biodetection systems must discriminate between unnatural events and naturally occurring backgrounds.

For all detection systems it is important to recognize that detectors alone do not directly measure threats to people. For instance, chemical detectors often cannot detect the lowest dose known to be a health risk without a preconcentration step, and biological detectors usually cannot determine viability of an agent or other important health-associated attributes, such as antibiotic resistance. Historically the military has set standards for chemical and biological detection for battlefield scenarios. The civilian community needs standards and uniform test protocols to ensure that systems perform predictably and that decision makers will be confident they will understand the information provided. Civilian applications demand high standards for performance, including low false alarm rates. An important role for the EPA Safe Buildings Program would be the establishing of these standards and uniform test protocols.

Detection to Assess Containment and for Post-Decontamination Evaluation

The stated focus of the Safe Buildings Program is building decontamination,[5] and it is in decontamination rather than detection that the EPA has leading expertise. Thus, the detection program should focus exclusively on decontamination, as indicated by the information provided to the panel from EPA. All technical parts of the program should be subordinate to decontamination and each part should technically support decontamination to achieve results within the prescribed time period of the program (approximately 30 additional months).

The specific components of detection[6] that the committee feels should be highlighted are

- the creation and publication of decontamination standards;
- specific specifications for detectors to support decontamination; and
- field-testing of detectors and decontamination strategies against the specifications provided above.

The committee concludes that EPA should lead the development of cleanup standards and sponsor and supervise realistic field tests to validate the decontamination protocols, including detection equipment for determining that necessary re-entry levels have been met. Beyond that, for the area of detection the measures of success given by EPA all fall into less important categories and should be removed from the EPA's funding portfolio since they do not support the program's primary objective.

Test-Beds and Protocols for Technological Systems

EPA's expertise in building safety has been on facility rehabilitation. As such, the concept of restoration of service is a key tenet of the EPA's core strength. To continue to effectively accomplish the mission of building safety and restoration EPA needs to focus its research in the Safe Buildings Program on decontamination and disposal. The components of the Safe Buildings program that are related to detection need to be fully directed toward support of the primary mission of decontamination and disposal. The components of the program that deal with detect-to-warn, in the

[5]EPA, 2003, Research Planning in the Safe Buildings Program—Report to the National Academies, presented to the committee on June 30, 2003, by N. Adams, Environmental Protection Agency, Research Triangle Park, NC, p. 5.

[6]EPA, 2003, Research Planning in the Safe Buildings Program—Report to the National Academies, presented to the committee on June 30, 2003, by N. Adams, Environmental Protection Agency, Research Triangle Park, NC, p. 23.

committee's view, have little possibility for benefit to either EPA or Homeland Security under the scope and duration of the program, as the program duration is too short to successfully complete this effort. Due to the limited time available, EPA should focus on decontamination. Any research on detect-to-warn, other than by collaborating with other agencies, is outside the current scope of EPA, and it would be unrealistic for EPA to expect any results in the remaining time.

A second overarching issue is the need to accomplish results within the program time frame. The committee detected a lack of focus during the first six months of research that has resulted in a dilution of the overall program. The committee feels that significant restructuring of detector research is needed, because detection has a number of components that are important to decontamination efforts. The committee feels that in the remaining 30 months in the program, the detection area should be focused to support the more tractable area of decontamination.

Specifically, EPA needs to focus a significant amount of its detection capabilities on answering the question "How clean is clean?" The establishment of standards for decontamination is within the purview of EPA, and it should take the lead in these efforts and be assisted by NIST and the Science and Technology Directorate of the Department of Homeland Security.

In summary, detector development should play a subordinate role to other components of EPA's Science and Technology program for homeland security. The area of detector development is not one in which EPA should play a leading role, nor is it one in which EPA should play a development role. EPA should be involved in setting standards for decontamination, as well as in testing commercial equipment to achieve the level of decontamination standards that detection will certify. This means that EPA must collaborate with other agencies to establish relevant standards and testing protocols.

As noted above, the committee recommends that EPA not pursue detect-to-warn systems further (due to the complexity of the issues and because this is outside EPA's primary expertise). If EPA chooses to pursue further research in this area, it is essential that the specifications of the warning system be tied to meaningful endpoints. For example, acute toxicology (or infectivity) thresholds for agents of concern should be considered, and the detection limit should be low enough (considering sampling time) that the agent can be detected before causing significant toxicity. If an agent can be detected before causing toxicity, then individuals that would have been infected otherwise have a chance to get out of the building and the decontamination procedure can begin at an earlier time.

Specifications for detection for re-entry should take into account the toxicology of the agents of interest. The detector should be able to detect levels exceeding chronic health guidelines (e.g., exceeding the reference concentration) following measurement for a specified period of time. Sampling time for reliable measurements should also be considered, although such considerations are less limiting, because measurement for re-entry can include sampling times on the order of days.

Role of the EPA Environmental Technology Verification Program

Establishment of required chemical and biological agent detectors and detection systems for a range of potential applications is an important challenge for EPA's Safe Buildings Program. More robust, sensitive, specific, and faster responding sensors will be necessary for successful containment, decontamination, and potentially even disposal activities. However, specification of sensor capabilities will not be sufficient; detection instruments and integrated sensor systems must be proven under realistically simulated conditions using both surrogate and, in many cases, actual chemical and biological substances.

Substantial technical effort and resources will be required to test threat agent detectors and detection systems. Technical skills will be needed in

- designing realistic tests;
- creating valid sampling and detection protocols;
- arranging accurate instrument calibration samples and variable-strength, statistically robust, contaminant challenges in appropriate matrixes; and
- accurately scoring the performance of a range of instruments based on disparate physical, chemical, or biological principles.

The difficulty increases when the use of military grade chemical or biological agents requires a level of isolation and containment available only in certified surety facilities. To be credible, verification testing must be done in independent trials designed, operated, and scored by third parties with no direct stake in their outcome.

Through its Environmental Technology Verification (ETV) program EPA has experience in certification testing of pollutant detectors and detection systems and would be appropriate to extend this program to cover threat agent detectors and detection systems. For current ETV projects an EPA contractor designs the test program and invites instrument developers and vendors to bring prototypes or early production instruments to the test for certification. Scientists and engineers from EPA and other agencies may contribute to test specifications and design and also attend the verification tests.

CONTAINMENT

Importance of Understanding Buildings

Buildings represent an important threat and exposure category for the deliberate or accidental release of chemical or biological agents. Most of us spend between 75 and 90 percent of our time in buildings at work, at home, in schools, or in commercial establishments. Buildings themselves and their occupants may be direct targets of a chemical or biological attack or an indirect target (downwind of a release targeted at another facility); in either event buildings can offer potential safe havens. Buildings and building operations can also help confine and spread harmful contaminants.

The focus of EPA's containment research is on the development and testing of methods to prevent the spread of contaminants thereby protecting building occupants, first responders, and decontamination crews. The overall objective is to reduce or eliminate the impact of a chemical or biological attack on building occupants as well as to provide techniques and guidance to determine the efficacy of chemical and biological protection measures for new and existing buildings. The research seeks to understand the building so as to aid in safe, efficient, and cost-effective restoration.

The spread of contaminants in a building is generally by air circulation in large open spaces, airflow between connecting spaces, and airflow in heating, ventilating, and air-conditioning ducts. To devise a containment strategy it is important to understand the physics of air exchange caused by mechanical means, such as HVAC fans, buoyancy, and external wind conditions.

There are many building types, and containment must take into account their characteristics. They range from large open-plan buildings, such as stadiums, shopping centers, and transportation hubs, to buildings subdivided into modular spaces, such as apartment buildings and offices with separate workspaces. Some buildings are designed as sealed boxes ventilated mechanically, while others are open to the outside air through operable windows and other large apertures.

Research is needed to understand the rate at which a contaminant introduced into a specific interior space spreads throughout the building. For contaminants introduced into the intake or ducts of an HVAC system, the spread and concentration levels throughout the building must be predicted. Exposure within the building from exterior sources must be tied to the performance of the building envelope and the HVAC system.

Understanding of contaminant distribution must then be coupled with an understanding of HVAC system operations and control systems in order to develop short-term protection strategies. The strategies range from isolation of an individual space to modified operation or shut down of HVAC circulation systems to building evacuation through spaces protected from contaminant intrusion. Protective measures using HVAC shutdown must be evaluated against adverse health effects caused by loss of ventilation over varying time periods.

A better understanding of building classes is needed to guide longer-term decontamination efforts as well. This will help define locations for monitoring and sampling based on predicted contaminant dispersal patterns. Ventilation techniques to clear the building of decontamination agents are linked to an understanding of airflow patterns under different HVAC operations. The building materials and furnishings may exhibit long-term emissions of contaminant agents and decontamination substances.

Relevance of EPA Program

EPA has a long history of research both on indoor air quality and on buildings. Some of this work has been intramural, while a significant amount has been extramural support for research at a variety of institutions, such as academia to the national laboratories and private industry and organizations. Containment of toxins, released either indoors or outdoors, so that human exposures are avoided or minimized, will clearly benefit from research on indoor air and on buildings. However, the containment research proposed by the EPA's National Homeland Security Research

Center (NHSRC) is more an assembly of existing research ideas than what is needed to answer critical questions in the field. An additional difficulty is the program's three-year limit (to the beginning of fiscal year 2006), which means that a longer-term research agenda cannot be formulated and work on key elements begun until the deadline is renegotiated or work is re-integrated into other EPA research centers and activities without loss of direction or commitment.

EPA appears ill-suited for a short-term research program in the area of "active" containment. In the time remaining EPA containment research should have three main objectives: (1) ensure that decontamination strategies account for all potential transport and fate pathways and surfaces in a building to ensure that building decontamination, cleanup, and disposal practices effectively and economically meet the desired criteria; (2) identify and prioritize key long-term building science and containment research needs and determine the role EPA should have in conducting or funding the research; and (3) in collaboration with industrial organizations and other agencies, develop design and performance criteria and standards for contaminant containment systems and methods.

Containment Focus

One area of containment covered by EPA's mandate that is of crucial importance to future EPA recommendations is the role of containment during decontamination and disposal. Containment and disposal will be handled by professionals who will be actively controlling the environment of the building. The kinetics of containment can be better understood in these controlled circumstances. However, people involved in decontamination and disposal operations will inevitably face problems of continued transport of chemical and biological agents, in part through resuspension of previously deposited particles. Understanding these problems and having supporting data and models to address these requirements should be an important, immediate EPA task, and could lead to guidelines in a shorter time frame than work on active containment. Areas of possible EPA focus are HVAC use, aerosol and gas penetration through doors, and inadvertent dispersal of contaminants (i.e., many of the themes to be studied by EPA in the broader containment framework) specifically in the framework of decontamination and disposal.

Another potential research topic is to use airflow and contaminant transport and fate modeling to help examine some of the contamination issues. It is an important task to protect the public by ensuring that decontamination materials are not lost to the environment and to maintain control over the building airflow to ensure that a building is decontaminated as efficiently as possible. Airflow circulation within a building can scatter contaminants and decontamination agents in ways that may limit the efficiency of the decontamination effort. Therefore, control is necessary to prevent contaminants from re-entering decontaminated areas and allow decontamination agents to properly disperse so that toxic materials can be neutralized.

Identify Long-Term Research Needs and Determine EPA's Role

Many of the containment issues cannot be addressed adequately in a three-year program, in part because of the dearth of past and current research in this field by EPA, other government agencies, and academia. The committee recommends that EPA use the next few years to define a comprehensive long-term research program in this area, and it should be carried out with other organizations knowledgeable in this field, including NIST, DOE, the Department of Homeland Security and academic researchers in the United States, Europe, and Japan. This research should incorporate the following:

- Evaluation of existing multizone airflow and transport models, such as CONTAM (NIST) and COMIS (LBNL), for predicting air and contaminant movement for application to the problem of understanding the transport and fate of biological and chemical contaminants and decontaminating agents;
- Development or evaluation of modeling methods for detailed simulation of contamination dispersal in a single space that can be coupled to a multizone model for overall building behavior to specifically include ductwork and vents;
- Development of containment strategies for important generic building types to include securing safe spaces, evacuation, and decontamination;
- Coupling of design of HVAC systems and controls with potential fast detection technologies for releases in and outside buildings;

- Prediction of contaminant dispersal patterns to aid in decontamination;
- Development of effective means for removing gaseous decontamination products;
- Characterization of the long-term reemission of contaminates from building materials and furnishings; and
- Development of a building taxonomy for categorizing new and existing building stock according to building operations, contaminant transport, and other characteristics.

EPA has begun to identify the long-term research needs in this area and whether existing or planned programs in other agencies effectively explore them. The agency has recognized that much of this research will be collaborative, either in funding specific projects with other governmental agencies or funding separate projects that have collaborative or synergistic outcomes. A key activity for which EPA is well positioned is to identify some common research goals and priorities across the various agencies, both as a way of directing EPA research to fill in critical research gaps and as a means establishing interagency cooperation in sharing knowledge, if not research goals.

Building Performance Criteria and Standards

Projects in the proposed containment Research Implementation Plan (RIP) cover a variety of fundamental and applied topics on the mitigation of chemical and biological warfare (CBW) attacks. The RIP does not consider one of the most important needs for containment system design: the development of design criteria and performance standards.

For example, a comfort air-conditioning system is designed to maintain a particular indoor temperature and humidity given specified occupancy and weather conditions. The indoor conditions are selected to ensure that at least 80 percent of the occupants of the conditioned space will be satisfied with the thermal environment. The parallel analysis for a CBW containment system could be to limit the maximum exposure over a defined period in a given space to a design-basis release of an agent. Considering the spectrum of CBW agents that could be used in a terrorist attack, a design-basis set of agents, both gas-phase and aerosol, may be needed.

The performance standards that must be met by a system or component and the design conditions that define the reasonable worst-case scenario under which the system must operate drive the design process. Without them the level of protection afforded by a system is not well defined. Consequently, the engineers and other professionals responsible for designing and implementing containment systems will not have all of the information they need to cost-effectively ensure the desired performance.

Successful pursuit of this objective will require collaboration with agencies outside EPA with high expertise in HVAC systems, air flow analysis, and health effects of chemical and biological exposures. To fully achieve consensus design standards, interaction with organizations such as American Society of Heating, Refrigerating, and Air-Conditioning Engineers and the American Society for Testing and Materials may be required. This is a longer-term project but should be initiated as soon as possible because of its importance. Even interim standards and criteria would be of great value.

The vast number of chemical and biological agents, each with its own toxicity signature, and the essentially unbound number of building types, creates a challenge to providing meaningful advice regarding containment during an attack. Conceivably, advice pertaining to HVAC design and operation, while relevant to the containment of chemical and biological agents in the event of contaminant entry into a building, might actually lead to the spread of the contaminant if acted on in either an untimely manner (i.e., after contaminant entry into the building) or in a building context not imagined by the agency in preparing the advice. The understanding of how contaminants spread in a building (even how they spread from the opening of an envelope in a very specific room scenario) will need to be gained through extensive experimentation, evaluation of data currently in existence, and generation of data in likely threat scenarios. This understanding will inevitably parallel to some degree the improvement of detection equipment; in any case, such knowledge will need to be acquired over a period of time far greater than the current EPA mandate.

DECONTAMINATION

A complete national-level program for protection against CBW agents and toxic industrial chemicals (TIC) must also include elements to respond to an attack. It is essential to develop and test technologies to mitigate and decontaminate the effects of a CBW agent or TIC attack on facilities and people as well as to restore critical facilities after they are attacked. This committee sees EPA's primary role as one of giving the nation an ability to completely restore domestic facilities rapidly and safely after an attack.

The committee agrees that the need to develop and implement effective decontamination technologies and health-protective cleanup standards for civilian scenarios is urgent. The anthrax incidents illustrate important problems associated with decontamination of civilian-sector facilities. For example, more than three months were required to clean up the Hart Senate Office Building at an EPA cost recently reported to be about $27 million, while the American Media, Inc., headquarters building in Florida, which was first to receive a letter containing a powdered form of anthrax, remains sealed and abandoned to this day. When considering how to address decontamination and cleanup in the public sector it is important to recognize that an effective response must be not only agent specific but site specific as well. The Safe Buildings Program is primarily aimed at indoor facilities such as an office building or hotel, where decontamination of ventilation systems is imperative, and public perception issues are extremely important. It also includes a semi-enclosed setting, such as a subway or transportation node. EPA has identified a list of high priority threat agents and the committee has considered those in its assessment below.

In general, the committee feels that the Safe Buildings Program is on target with respect to ongoing and proposed decontamination efforts. It is recognized that a great deal of effort went into making sure that the program was coordinated and not duplicative with ongoing efforts of other agencies and the Department of Energy national laboratories. It is also recognized that this is not an easy task and that some agencies continue to be less than forthcoming with regard to the sharing of information. In the limited time available the Safe Buildings Program has identified some key areas that are not currently being pursued by other agencies but fall in the EPA's jurisdiction and expertise. EPA has focused its efforts on four areas, and the committee provides more specific comments for each area below. These efforts should be coordinated with the disposal task to ensure that the resulting waste and by-products are optimized for disposal. Additionally, the committee suggests that more effort and resources be expended on developing an extramural research program that goes beyond current contractors. Academic laboratories must be encouraged to participate in longer-term research.

Development of Standardized Test Protocols for Decontamination Technology Performance

The development of standardized test protocols for decontamination technology performance is an important area for EPA to pursue. In the United States new formulations sold as a biocide must meet EPA approval under the Federal Insecticide, Fungicide, and Rodenticide Act (FIFRA). The committee agrees that current testing requirements need to be changed because they are obstacles to the commercialization of new bio-decontamination formulations, many of which are based on oxidizer systems for which the tests were not designed. EPA is aware of this and has formed an interagency committee to design more rapid and accurate testing protocols. This activity needs to be accelerated and needs to investigate all types of potential biocides, whether they exist as solutions, foams, gels, gases, or aerosols.

One major additional need that should be addressed is the sampling and analysis protocols to be used for cleanup of facilities following a CW or BW incident. Issues involving spore recovery from porous surfaces have not been adequately addressed and could have a major impact on the interpretation of results. Quenching should be better explored and standardized since the use of surfactants in a formulation can also impact the consistency of biocide results. EPA has procedures for CW and TIC that are adequate but should be further evaluated for the threat agents that EPA has identified. Some of the sampling and analysis procedures developed in support of the Organization for the Prohibition of Chemical Weapons may be easily implemented for this application.

Overall EPA focus should be to develop written standards and protocols that distinguish clearly between requirements for chemical and toxic industrial chemicals and biological threats. Additional guidance and statistical sampling protocols should be developed similar to those developed and used by DOE and EPA for radiological sampling (EPA et al., 1997).

Systematic Study and Verification of Decontamination System Performance

The committee supports the ETV proposed in principle and feels that this is an important focus area for the EPA's Safe Buildings Program. To be effective, EPA must build off the lessons learned from the anthrax incidents as well as other studies that have been done at Dugway Proving Ground (Larsen, et al., 2002) and elsewhere (Raber and McGuire, 2002) within the last three years. The focus for this evaluation should be on civilian sector issues, where decontamination requirements are extremely demanding and different from requirements for the military. Military testing protocols are aimed at formulations that decontaminate in 30 minutes or less and have not focused on many of the environmental issues facing public sector use. Few controlled protocols exist for testing various building materials, and there is a limited experience base for testing gases, especially in ventilation system decontamination. In addition, the potential for resuspension of previously deposited particles must be better understood to allow for optimizing decontamination. The indoor air-modeling discussion in the "Containment" section addresses this in more detail.

Key test attributes must be determined and weighted by stakeholder group; here again they differ from typical military requirements. What is desired is a method that is noncorrosive, nonhazardous, and degrades to environmentally acceptable residues (i.e., passes EPA methods 8260 and 8270 for volatiles and semivolatiles) and one that works over relatively short decontamination times (hours). Public sector applications also require a formulation that makes maximum contact with surfaces by adhering to walls and ceilings but does not form toxic products by reacting with walls and ceilings; is relatively inexpensive and available; has a long shelf life (at least a year); and is easy to deploy and implement. In short, the public sector would be well served with a formulation that takes two days to decontaminate if the material were nontoxic or degraded to environmentally acceptable by-products. This would not meet military needs.

The committee is concerned about the number of products appearing on the market with decontamination claims. EPA is in a good position to evaluate these claims through an effectively structured ETV program. The committee feels that this is an important national role for EPA. Additionally, EPA needs to continue its collaborative working relationship with the Department of Homeland Security with regard to ongoing and planned demonstration programs for decontamination and restoration.

Evaluation of the Toxicology of Decontaminating Agents

EPA has correctly concluded that there is a need for information on the fate and safe exposure levels of the decontaminating agents themselves, since residues may linger after the building has been decontaminated, and it is important to determine whether it is safe to enter the building and occupy it. More information is needed about the rationale for the choice of compound being studied and for the experimental protocol; based on the available information, the proposed study appears unlikely to meet its desired objectives.

Of the possible gas-phase decontaminating agents (e.g., chlorine dioxide, vapor-phase hydrogen peroxide, para-formaldehyde, and ethylene oxide), EPA has chosen to conduct additional studies only on chlorine dioxide. No rationale for this choice was expressed in the RIP, but it appears to be because chlorine dioxide was used at the Hart Senate Office Building and because it is one of the agents of choice for anthrax. If chlorine dioxide is indeed the agent of choice for anthrax, and if anthrax contamination is among the most likely of the threat scenarios, then additional research on chlorine dioxide is reasonable. On the other hand, if there are other decontaminating agents with similar or higher efficacy than chlorine dioxide, it may be preferable to conduct additional research on these agents (see below).

It is not clear which data gaps the proposed studies will consider or how the data would be used in the development of building re-entry guidelines. Before addressing the available data on chlorine dioxide, the relevant exposure standard for re-entry should be considered. The investigators cite the existing NIOSH/OSHA recommended exposure limit. This is useful as an initial comparison, since the exposure of interest occurs in the workplace, and OSHA is the regulatory agency for workplace exposure. However, the exposure of interest would most likely occur primarily among office workers returning to the building. Although this is not a well-defined area of occupational risk analysis, office workers tend to have a lower acceptance of risk than workers in factories that are exposed to industrial chemicals. Office workers are likely to resemble the general population more than factory workers; the lower level of physical activity and the absence of selection for healthy people may remove much of the "healthy worker" effect. Although

EPA does not have jurisdiction over the office environment, evaluation of the risk of long-term exposure for office workers should take into account health standards developed for the general public (e.g., the reference concentration [RfC]). The RfC is defined as "an estimate (with uncertainty spanning perhaps an order of magnitude) of a continuous inhalation exposure to the human population (including sensitive subgroups) that is likely to be without an appreciable risk of deleterious effects during a lifetime" (EPA, 2000).

A second exposure of interest is workers reentering a building to complete cleanup and replace furnishings. In this case, exposure would be for a limited duration, and short-term guidance values (e.g., for an eight-hour or one-week exposure) would be appropriate. If there are data gaps in this area, the proposed studies may help in evaluating the toxicity of chlorine dioxide. However, because the proposed high concentration of 0.1 ppm is near the "no observed adverse effect" level (NOAEL) for sub-chronic exposure, it is not clear whether the proposed exposure concentrations would provide the necessary information. Higher exposure levels may be needed to identify a NOAEL and a "lowest observed adverse effect" level (LOAEL) for acute exposures. EPA assessed the inhalation toxicity of chlorine dioxide in 2000, and has derived an RfC for this chemical (EPA, 2000). While the RfC is significantly lower than the REL, due to differences in the methods used, the documentation for the RfC identifies key data gaps and uncertainties that should be considered in conducting further testing. In particular, there is uncertainty regarding the NOAEL in the subchronic study used as the basis for the RfC. Additional short-term mechanistic studies will not resolve that issue. Another key data gap is the absence of studies of reproductive toxicity through the inhalation route. While gene expression studies are useful for hypothesis generation, they will not provide definitive answers regarding the carcinogenic potential of chlorine dioxide, particularly in the absence of a chronic cancer bioassay.

Based on these considerations, the proposed studies may provide useful general information in chlorine dioxide toxicity, but they are unlikely to provide information within the time frame of interest that is useful to evaluating the health risks of chlorine dioxide following re-entry. More relevant information would be obtained by targeting the testing conditions to determine the NOAEL and LOAEL for the exposure duration of interest (acute or sub-chronic). These standard toxicity studies could be supplemented by mechanistic studies, such as studies designed to improve dosimetry or to establish the mode of action. However, the mechanistic studies should not be the focus of the testing, in light of the objectives of the overall Safe Buildings Program.

In summary, the committee recommends that the EPA Safe Buildings Program concentrate on the potential for generation of toxic residues from the interactions of decontamination agents with building materials, furniture, carpets, and other furnishings and cover more than just chlorine dioxide. Understanding the fate of the decontaminants with regard to specific building materials may be useful, but more important is the actual resulting residues and whether they represent a significant health risk.

Engineering and Scientific Support and Analysis

EPA has identified an initial set of projects to go forward with in this area. The key issue for the EPA Safe Buildings Program will be to come up with the necessary information and supporting processes to be able to make risk-informed decisions when mounting a response. Any kind of decontamination in the public sector requires an understanding of the type of emergency restoration needed, including characterization and performance of site-specific risk assessments for potential impact on human and animal health and the environment. These considerations then determine the decontamination or remediation treatment to be implemented (Raber et al, 2001).

The committee supports the current efforts of EPA to learn from ongoing anthrax fumigations and recommends that efforts should include a thorough case study of the Hart Senate Office Building and Brentwood Post Office experiences, focusing on decontamination and disposal. Understanding the real-world conditions and results should provide a baseline for engineering and economic analysis of building decontamination alternatives. Some of the more recent reports from EPA (2002) and the GAO (2003a,b) also provide a useful perspective and should be incorporated into this effort.

"How clean is clean enough?" or put another way, "How clean is safe?" The challenge is to establish target levels of cleanup for the various biological and chemical agents that will meet both regulatory and stakeholder needs and address site-specific parameters as well.

The committee feels that two concepts are essential to any discussion of cleanup levels. First, public perceptions and stakeholder issues will drive cleanup requirements. Second, economic drivers and inconvenience influence stakeholders to accept higher risks. Input from public and regulatory stakeholder bodies is essential when cleanup goals relating to health and risks are set, and when cleanup and decontamination decisions are made. The committee believes that successful cleanup requires that risk information be communicated to the public throughout the entire process. Moreover, the decontamination method that is selected needs to consider the costs of cleanup versus the goal of meeting cleanup criteria.

Following the anthrax letter incidents in the United States, the importance of economic drivers became clear when cleanup efforts commenced. Although regulatory guidelines and recommendations for specific airborne and soil cleanup levels for the G agents (tabun, soman, sarin, and cyclohexyl methylphosphonofluridate) and VX exist (U.S. Army, 1999; NRC, 1999b, EPA 2000), cleanup levels for biological agents remain problematic (Raber et al, 2001). To address standards and policies for decontaminating public facilities affected by exposure to harmful biological agents, in particular, anthrax and smallpox, the Department of Homeland Security has commissioned a new study by the National Research Council. The EPA's Safe Buildings Program is providing useful direction regarding the scope and direction of that study. The NRC committee encourages EPA to continue to work on this as an important interagency effort.

The committee also recommends that the EPA Safe Buildings Program consider adding the scope suggested below to its proposed research and implementation program.

High on the list of important issues is the need for methods to rapidly determine agent-specific viability for effective biological agent decontamination. Current methods, such as PCR technology, tell us what an agent is but not whether it is alive or dead. Current sampling and culturing methods take from one to three days, depending on the agent. Rapid determination of specific agents is the key to restoring critical infrastructure. Some initial research is underway at DOE laboratories, primarily funded by the Defense Advanced Research Projects Agency, to explore the technology gaps related to viability determinations, but this area is not likely to meet all of EPA's needs. The critical advantage of such a method needs to be integrated into EPA's current approach for biological decontamination.

It is important to understand the potential for natural attenuation for both chemical and biological warfare agents. It is well known that such factors as ultraviolet light from the sun can kill vegetative cells and certain other biologicals. Some literature is available for spore and cell survival in outdoor environments, but information is limited for indoor environments, and controlled environmental studies are lacking (Setlow, 2000). Research on *Bacillus anthracis* shows that the spores suffer heavy mortality over time—up to 90 percent per year—but remaining spores can germinate and grow. Nonsporulating vegetative cells that require high water activity survive longer in a dormant state. Thus, it becomes important to study the conditions required for germination, growth, and sporulation in enclosed and semi-enclosed environments, especially as a function of temperature, humidity, and time. Again, understanding the limits of natural degradation potential can aid in determining effective decontamination strategies.

With regard to CW agents, early work done by Lawrence Livermore National Laboratory researchers in cooperation with British scientists at Porton Down, U.K., studied natural degradation of the G agents and VX at a concentration of 5 g/m^2. Tests were conducted in sandy soil, silty soil, and on silicon rubber gasket material. Within three days in a humid environment the chemical agents degraded to nondetectable levels, except for the tests on rubber gasket material (Raber et al., 2001). Thus, the expectation is that outdoors and at least in soil, natural degradation of chemical warfare agents can be effective. Understanding this for semi-enclosed environments may also aid in determining effective decontamination strategies.

Long-Term Research Needs and EPA's Role

The EPA Safe Buildings Program needs to help set the stage for a longer-term research agenda for decontamination and restoration. EPA has identified three areas that are required for longer-term success. These are (1) development and optimization of novel (and improved) decontamination methods; (2) additional evaluations of existing decontamination methods and systems; and (3) development of methods for high-value materials (e.g., museum holdings, national treasures).

The committee agrees with EPA's longer-term recommendations. The committee evaluated current technologies to identify significant decontamination technology gaps for applications in the public sector. The committee recommends that a longer-term decontamination and restoration research program, coordinated with other agencies, include the following:

1. Technologies, systems, and studies to better characterize the extent of chemical and biological contamination resulting from a terrorist attack using CBW agents or TICs.
   - **Standards Development**—studies to determine standards for cleanup levels for CBW and TIC materials for various types of contaminated facilities and demographic groups.
   - **Sampling Methodology Development**—technologies to increase sampling efficiency for CBW agents in facilities and surrounding areas.
   - **Agent and Decontaminate Studies**—studies to develop a better understanding of transport, robustness, and viability of CBW agents and the potential for toxic residues resulting from the decontamination process.

2. Development of methods for better, cheaper, safer, and faster decontamination of facilities and other contaminated areas.
   - **Decontamination of Sensitive Equipment and Other Items**—technologies and methods to decontaminate other sensitive equipment and sensitive items (e.g., computing equipment, paintings).
   - **Decontamination of Hard-to-Reach Places**—technologies and methods to decontaminate hard-to-reach places such as the interior of ductwork and the area above ceiling tiles. These technologies should include less toxic vapor and gas decontaminants and methods that can be employed and utilized with less infrastructure requirements than chlorine dioxide.
   - **Decontamination of Exposed Surfaces and Wide Areas**—technologies and methods to effectively decontaminate exposed surfaces in facilities and surrounding areas (i.e., specialized reactive and sealant materials in paints or coatings that could adhere to high places and require no cleanup and methods to decontaminate; foam, gels and liquids that adhere to vertical surfaces and ceilings.
   - **Biological Mechanistic Decontamination Studies**—studies to determine mechanisms of spore kill and other related effects.

3. Restoration Systems Development and Evaluation—the study and development of systems and tools to better understand, to plan, and to implement the restoration process (e.g., identification of preventative measures for sealing and covering porous surfaces with materials, such as epoxy paint or stainless steel sheeting, to create smooth surfaces that are easier to decontaminate). This includes the verification of decontamination products and the optimization of the process to be used.

It is clear from the efforts under way to use chlorine dioxide and vaporous hydrogen peroxide that there is much scientific data lacking (e.g., effectiveness as a function of temperature, humidity, time, materials) that is vital to an efficient and effective decontamination. Currently, huge infrastructure needs (e.g., source generation, air-tight tenting of building, negative pressure apparatus, circulation pumps, scrubbers, detection apparatus) are required and this causes a significant delay in the restoration of contaminated facilities. A proper metric for this program would be to decontaminate and restore a building similar to the Hart Senate Office Building in less than one month and preferably two weeks.

DISPOSAL

The proposed strategy and research projects reflect EPA's expertise in handling hazardous materials disposal issues, its experience in responding to the recent anthrax decontamination effort, and its overall experience in hazardous waste cleanup through implementation of Superfund cleanup and removal actions.

Specific research and implementation projects focus on disposal options that include thermal or incineration and landfill. The grass-roots approach EPA is taking to involve impacted regulatory agencies, manufacturers, emergency responders, and facility operators is a sound approach. Both thermal treatment and incineration have the advantage of addressing both hazard and waste reduction issues. The need to understand the matrix effects of incinerating bulk items to destroy CBW agents is a necessary step toward using this technology. Developing methods to test incinerator emissions for the agents being treated is essential for effectively using thermal treatments for CBW agents and meeting public perception issues and stakeholder demands. Looking at the viability and survivability of organisms of concern in landfills is also necessary to allow for disposal of BW materials directly into landfills. Research focused on the longevity of *Bacillus anthracis* spores shows that 90 percent of spores in soil die within 50 years (Sneath, 1962). However, surviving spores can remain viable for 300 years. Understanding this issue and the associated impact is key to an effective disposal strategy.

This committee supports EPA's approach to the proposed projects on disposal. However, it also recommends that EPA consider the following in its CBW disposal options:

1. Thermal treatment and incineration may not be viable approaches in many states where air quality issues are a concern. For example, California does not have a permitted hazardous or medical waste incinerator and only has three municipal facilities that use incineration as a permitted waste transformation process.

   There are only a handful of operating incinerators around the country. Investigations as to whether these few incinerators can handle the bulky items identified in the project proposals need to be included in EPA's consideration.

   The project proposals also include streamlining the permitting process for thermal treatments. However, they do not include issues related to whether there are technologies that can economically meet air quality standards and whether public perception will allow such facilities to be sited even if the permitting process were streamlined.

   Thermal treatment temperatures that effectively kill any remaining anthrax spores are typically higher than those for standard medical waste incineration.[7] To render biological warfare agents completely harmless, dry heat requires two hours of treatment at 160 degrees C. If steam is used at 121°C and 1 atm of overpressure (15 psi), the time may be reduced to 20 minutes, depending on volume (Office of the Surgeon General, 1997). Therefore, these incineration and thermal treatment proposals should include a focus on the efficiency of different temperature treatments for the specific CBW agents.

2. A criterion needs to be developed for disposing of decontaminated materials in municipal landfills, which would allow more options with respect to facilities able to take decontaminated CBW waste. For example, in California, if criteria were available to determine that wastes are no different than any other municipal waste, including waste contaminated with sewage, CBW wastes could be disposed in an appropriate Class III landfill.[8,9] The key technical question for EPA to answer is whether the decontaminated material meets the criteria of any other municipal wastes. Answering this question positively could set baseline criteria for a decontamination standard and should be a key factor in setting goals and objectives for EPA's decontamination research program discussed in EPA's Research Implementation Plan.

3. EPA also identifies thermal treatments as a disposal option; however thermal treatments are also decontamination methods. In fact, the projects seem to be proposing thermal treatment as a way of treating residual contamination before disposing the waste products to air and landfill. The decontamination and disposal strategies and proposals should be better coordinated. If the material is to be decontaminated, then decontamination methods that will meet municipal landfill criteria as discussed above should be identified.

4. The committee recommends that EPA have a strategy related to methods development for stabilizing remaining hazardous waste materials so that the materials may be sent to a landfill rather than impose an incineration or thermal treatment approach. Such items as air handling filters and other more porous materials may be more difficult to decontaminate. It will be important to evaluate

---

[7]Title 17, California Code of Regulations, Division 1, Air Resources Board, Chapter 1 Air Resources Board, Subchapter 7.5. Airborne Toxic Control Measures, Section 93 104, Dioxins Airborne Toxic Control Measure—Medical Waste Incinerators

[8]Title 27, California Code of Regulations, Environmental Protection Division 2, Solid Waste

[9]Title 14, California Code of Regulations, Natural Resources Division 7, California Integrated Waste Management Board pertaining to nonhazardous waste in management in California.

some of the more recently developed resin stabilization methods for nuclear-contaminated material in the context of the Safe Buildings Program. Some such methods may stabilize remaining spores and be applicable to treatment for toxic industrial chemicals, including CW materials. The committee encourages EPA to discuss some of the approaches used by the DOE national laboratories where they have been successful in stabilizing difficult wastes to meet land disposal restrictions and dispose of the wastes directly in a landfill (Tyson and Schwendiman, 1995; Bowers et al., 1995; Gates-Andersen et al., 2003).

5. The EPA's proposals discuss liquid wastes generated as part of decontamination efforts, but they do not include research regarding disposal of this waste or whether there is a need to address liquid waste disposal options. For example, can these types of waste be discharged to a publicly owned treatment works (POTW) without interrupting treatment processes or causing exposures to workers at the POTW. In some cases the decontaminated wastewater might be characterized as hazardous for toxicity. It is important to determine whether the current hazardous waste disposal methods are indeed adequate for handling the liquid waste stream created by decontaminating CBW-contaminated materials. Approaches should be evaluated for stabilizing any remaining hazardous liquid wastes that are generated so that they can meet landfill toxicity characteristic leaching procedure (TCLP) disposal criteria. Such approaches should be incorporated into EPA's overall strategy.

In summary, EPA should develop guidance documents and matrixes for emergency managers on what the best disposal options would be as a function of potential impacts on environmental media (e.g., air, water, land) receiving the different types of waste material. EPA also needs to include an analysis of competing and overlapping federal, state, and local regulations and ordinances that might prevent the application of certain disposal technologies. This would be the key to streamlining disposal procedures.

3

Findings and Recommendations

The committee's overall findings and recommendations are listed below. The high-level findings and recommendations are listed first and represent what the committee deems most important. These are followed by more general findings and recommendations, and finally, comments are included specific to major research areas. Detailed recommendations are found in Chapter 2. As noted earlier, the committee chose to provide EPA review and recommendations for key areas rather than perform a detailed review of every proposal in the RIP. Given the short time frame, the committee felt this would best provide EPA guidance for the remaining time in the program.

OVERARCHING FINDINGS AND RECOMMENDATIONS

EPA has correctly identified the major research areas essential for the Safe Buildings Program.

The primary areas of research associated with an effective building decontamination strategy are presented in the Safe Buildings Program Research Implementation Plan: detection, containment, decontamination, and disposal.

The duration of the current program is insufficient to deal with all the tasks and goals presented in the Research Implementation Plan.

As specifically noted in the various findings and recommendations, the program time frame (scheduled to end at the end of fiscal year 2005) is too short to effectively accomplish all the goals set forth in the Research Implementation Plan. The proposed plan covers an extensive area of new research, of which the scope and breadth is too large to accomplish in the allotted time frame. Given current resources and the extramural collaborations of EPA, it is unrealistic to expect results in all areas of proposed research in the remaining time.

The current effort should include a planning function for a potential longer-term research program to address unmet needs in technical areas.

The short time frame proposed in the Research Implementation Plan is an overarching concern; accordingly, the committee has made specific recommendations for tasks that can reasonably be completed in three years or less. One of these tasks should be the development of a coordinated program for a long-term research and development effort focusing on the agency's strengths. In some areas, the committee has provided guidance or recommendations for long-term research activities.

In the short term, the program should focus almost exclusively on decontamination and disposal issues

and other parts of the program should be subordinate to decontamination and disposal.

EPA has expertise in the area of decontamination and disposal. The agency should focus the remaining time toward improving these aspects of its work. Activities in the remaining areas—detection and containment—should technically support decontamination and disposal in a logical manner to achieve results within the prescribed time period of the program. The committee has made specific recommendations for these areas, as described below.

GENERAL FINDINGS AND RECOMMENDATIONS

EPA should develop a more interactive system to communicate and collaborate with other government agencies.

Throughout the course of its review, committee members were able to identify related programs and projects in other agencies that were not accounted for in EPA's planning process. This is understandable given the short time EPA had to start up the Safe Buildings Program. More effort and resources should be expended in communicating with other agencies having research programs in these areas. Academic laboratories should be encouraged to participate in longer-term research. Participants should include people from other government agencies, industry, and academia. EPA should develop an external scientific advisory board for this program that does not include current contractors.

The key research areas require milestones toward meeting the goals of the program and measures of success that concretely reflect the degree to which EPA's recommendations improve building safety in accordance with program goals.

The Safe Buildings Program will likely be most successful if specific, realistic program goals are set early in the program and success or failure is systematically measured against those goals as the program evolves. This will ensure that each program area is adequately progressing as well as ensure that resources remain focused on successful ventures.

SPECIFIC FINDINGS AND RECOMMENDATIONS

Detection

In the short term, EPA does not have the capabilities to develop new preventative detection technologies.

Very specific technical skills are required for the difficult tasks associated with developing preventative (i.e., "detect-to-warn") detection technologies. The tasks include designing realistic tests, creating valid sampling and detection protocols, arranging accurate instrument calibration samples and variable strength, statistically robust, contaminant challenges in appropriate matrixes, and accurately scoring the performance of a range of instruments based on disparate physical, chemical, and/or biological principles. Because this is a relatively new activity for EPA, the design of detection technologies for use in real-time warning or reaction modes is not realistic given the three-year time frame of the Research Implementation Plan.

EPA's role in detection should be to set standards for required detector performance (e.g., limits of detection, accuracy, detection time, operational availability, etc.)—either for agents or for decontaminants.

There are three detection areas for which EPA should play a role. First, detection for determining that re-entry of a building is safe is a valid research area and within the scope of EPA's authority and expertise. Second, EPA should also be involved in setting standards or requirements for detection limits, particularly through interagency coordination. Third, EPA has experience in certification testing of pollutant detectors and detection systems through its Environmental Technology Verification (ETV) program. The agency can thus play a role in setting the test protocol standards and sponsoring the test beds for testing commercial equipment to meet the detection standards provided by EPA.

Containment

In the short term, EPA does not have the expertise to develop active, real-time building protection systems.

The challenge of providing meaningful advice related to real-time containment of a biological or chemical attack lies in the vast number of chemical or biological agents, each with its own toxicity level, detection signatures, and dispersion characteristics. In addition, there is an essentially unbounded number of building types. Conceivably, advice pertaining to HVAC design and operation, while relevant to the containment of chemical and biological agents in the event of contaminant entry into a building, might actually lead to the spread of the contaminant if acted on in either an untimely manner (i.e., after contaminant entry into the building) or in a building context not imagined by the agency. In addition, the proposed containment research appears to be an assembly of existing research ideas rather than a response designed to answer critical questions in the field. A longer term research plan is needed to effectively address these issues.

The RIP should differentiate between long-term versus short-term goals in the area of containment.

The committee recommends that EPA use the next few years to define a comprehensive long-term research program in this area. This should be carried out in conjunction with other organizations knowledgeable in this field, including National Institute of Standards and Technology, Department of Energy, Department of Homeland Security, and academic researchers. Detailed recommendations for long-term research goals can be found in Chapter 2.

The RIP should seek to develop building performance criteria and standards.

Projects in the proposed containment RIP cover a variety of fundamental and applied topics relevant to the mitigation of CBR attacks. However, the RIP does not address one of the most important practical needs for new containment system design: the development of design criteria and performance standards.

The containment focus in the RIP should be on decontamination and disposal efforts.

One area of containment that is in the purview of EPA and is of crucial importance to future EPA recommendations is the role of containment during decontamination and disposal. After contamination of a building by chemical or biological substances, those involved in decontamination and disposal operations will inevitably face problems of continued transport of chemical and biological agents, in part through resuspension of previously deposited particles. Understanding these problems could be an important task for EPA, and could lead to practical guidelines. For detailed recommendations on the decontamination effort refer to Chapter 2.

Decontamination

EPA has expertise and resources in decontamination technology and assessment.

In general, the committee feels that the EPA's Safe Buildings Program is on target with respect to decontamination efforts ongoing and proposed. The research program does need to be expanded to include gas phase decontaminants other than chlorine dioxide.

The RIP should fund efforts to develop standards and protocols for decontamination that distinguish between requirements for chemical and biological threats as well as develop and evaluate sampling technologies and protocols for decontamination.

The overall focus should be to develop standards and protocols that distinguish between requirements for chemical or toxic industrial chemicals and biological threats. Additional guidance and statistical sampling protocols should also be developed similar to those developed and used by DOE and EPA for radiological sampling (EPA et al., 1997).

EPA's role in decontamination should include evaluation of decontamination technologies through an effectively structured ETV program.

This is an important national role for EPA.

The EPA should help set a longer-term research agenda for decontamination and restoration.

Research needs include (1) development of methods to rapidly determine agent-specific viability for effective verification of biological agent decontamination, (2) technologies, systems, and studies to better characterize the extent of chemical and biological contamination resulting from a terrorist attack, and (3) development of methods for better, cheaper, safer and faster decontamination of facilities and other contaminated areas.

Disposal

EPA has a great deal of knowledge and background in the area of waste disposal.

The proposed waste disposal strategy and research projects therein reflect the EPA's expertise in handling hazardous materials disposal issues, its experience in responding to the recent anthrax decontamination effort, and its overall experience in hazardous waste cleanup through implementation of Superfund cleanup and removal actions.

In the area of disposal the RIP should concentrate on the interactions, effects, and potential for generation of toxic residues from the interactions of biological decontamination agents with building materials, furniture, and carpets.

Understanding the fate of the decontaminants with regard to specific building materials may be useful, but more important are the resulting residues and whether they represent a significant health risk.

EPA needs to conduct an analysis of competing and overlapping federal, state, and local regulations and ordinances that might prevent the application of certain disposal technologies.

Thermal treatments and incineration may not be viable approaches in many states where air quality issues are a concern. For example, California does not have a permitted hazardous or medical waste incinerator and only has three municipal facilities that use incineration as a permitted waste transformation processes.

There are only a handful of operating incinerators around the country. Investigations as to whether these few incinerators can handle the bulky items identified in the project proposals need to be considered by EPA. In addition, guidance documents need to be developed for emergency managers that specify the best disposal options available.

A criterion needs to be developed by EPA to determine the efficiency of decontamination efforts that would allow decontaminated materials to be disposed in municipal landfills.

Better understanding of the survivability of organisms of concern may allow for landfill disposal options. Additionally, the committee recommends that EPA have a strategy for developing methods for stabilizing waste materials so that they meet land disposal requirements.

Cross-Cutting

The decontamination and disposal strategies and proposals presented in the RIP are not well coordinated.

EPA identifies thermal treatments as a disposal option however thermal treatments are also decontamination methods. To be effective these overlapping areas need to be better coordinated.

The EPA's effort should include a case study of the recent anthrax decontamination experience, focusing on decontamination and disposal.

Evaluations of the responses to recent anthrax contamination of congressional and Post Office facilities demonstrate the severe management and technical difficulties posed by even a modestly scaled biological attack. Understanding the real-world conditions and results should provide a baseline for the engineering and economic analyses of building decontamination alternatives. A successful EPA Safe Buildings Program should show a significant decrease in the amount of time required to restore a building. The goal should be on the order of weeks and not months.

EPA should establish evaluation parameters, specifications, and needs for safe building systems and systems components and work with systems and components developers to establish their market potential.

For systems and systems components with significant market potential, EPA should use the ETV program to establish which systems and components meet required safe building specifications. This applies to decontamination, containment, and detection systems.

References

Bowers, J. S., J. R. Anson, S. M. Painter, and R. E. Maitino. 1995. Stabilization of inorganic mixed waste to pass the TCLP and STLC tests using clay and ph insensitive additives. In Proceedings of the Third Annual ASME Mixed Waste Symposium, Baltimore, Md., August.

CDC/NIOSH. 2002. Guidance for Protecting Building Environments from Airborne Chemical, Biological, or Radiological Attacks. DHHS (NIOSH) publication no. 2002-139. Washington, D.C.: Centers for Disease Control and Preventions and the National Institute for Occupational Safety and Health.

Cole, S. 2003. Biological/chemical sensors tested at San Francisco International Airport. Homeland Security Solutions Magazine 2(1):9-12.

Colton, R. J., and J. N. Russel, Jr. 2003. Making the world a safer place. Science 299:1324-1325.

EPA. 2000. Toxicological Review of Chlorine Dioxide and Chlorite, CAS No. 10049-04-4 and 7758-19-2.

EPA. 2002. Challenges Faced During the Environmental Protection Agency's Response to Anthrax and Recommendations for Enhancing Response Capabilities, A Lessons Learned Report. Washington, D.C.: U.S. Environmental Protection Agency.

EPA, DOE, NRC, and DOD. 1997. Multi-Agency Radiation Survey and Site Investigation Manual (MAARSSM). Washington, D.C.: U.S. Environmental Protection Agency.

GAO (U. S. General Accounting Office). 2003a. U.S. Postal Service. Issues Associated with Anthrax Testing at the Wallingford Facility. Report GAO-03-787T. May 19. Washington, D.C.: General Accounting Office.

GAO. 2003b. Capitol Hill Anthrax Incident. EPA's Cleanup Was Successful; Opportunitites Exist to Enhance Contract Oversight. Publication GAO-03-686, (June). Washington, D.C.: General Accounting Office.

Gates-Anderson, D.D., Kidd, S.D., Bowers, J.S., Attergery, R.W. March 25, 2003.HEPA Filter Encapsulation. U.S. Patent No. 6,537,350.

Larson, L., Harper, B., Rome, W., Ramachandran, C., Westwood, S. 2002. Abbreviated Test Report for the Laboratory Validation of Chlorine Dioxide Decontamination. West Desert Test Center, Dugway Proving Ground, Utah. (Available from the U.S. EPA, 999 18th Street, Suite 300, 8EPR-ER, Denver, CO, 80202.)

National Advisory Committee. 2000. Acute Exposure Guideline Levels (AEGLs) for Nerve Agents GA, GB, GD, and GF (CAS Reg. Nos. 77-81-6, 107-4-8, 96-64-0, and 329-99-7). NAC Draft 1, July. (Paul S. Tobin, designated federal officer.) Washington, D.C.: Environmental Protection Agency.

NRC (National Research Council). 1999a. Chemical and Biological Terrorism, pp. 55-180. Washington, D.C.: National Academy Press.

NRC (National Research Council). 1999b. Review of the U.S. Army's Health Risk Assessments for Oral Exposure to Six Chemical-Warfare Agents. Washington, D.C.: National Academy Press.

Office of Solid Waste (OSW). Method 1311, Toxicity characteristic leaching procedure. In SW-846, Test methods for evaluating solid wastes physical/chemical methods. Washington, D.C.: Environmental Protection Agency.

Office of the Surgeon General. 1997. Medical aspects of chemical and biological warfare. In Textbook of Military Medicine, part 1. Washington, D.C.: Department of the Army.

Raber, E., A. Jin, K. Noonan, R. McGuire, and R. D. Kirvel. 2001. Decontamination issues for chemical and biological warfare agents: how clean is clean enough? International Journal of Environmental Health Research 11:128-48.

Raber, E., and R. McGuire. 2002. Oxidative decontamination of chemical and biological warfare agents using L-Gel. Journal of Hazardous Materials B93:339-52.

Raber, E., J. M. Hirabayashi, S. P. Mancieri, A. L. Jin, K. J. Folks, T. M. Carlsen, and P. Estacio. 2002. Chemical and biological agent incident response and decision process for civilian and public sector facilities. Risk Analysis 22(2):195-202.

Setlow, P. 2000. Resistance of bacterial spores. In Bacterial Stress Responses, eds. G. Storz, R. Hengge-Aronis, pp. 217-230. Washington, D.C.: ASM Press.

Sneath, P. H. A. 1962. Longevity of microorganisms. Nature 195:643.

Tyson, D. R., and G. L. Schwendiman. 1995. Treatability studies involving epoxy solidification for various mixed wastes at Idaho National Engineering Laboratory. In Proceedings of the Third Biennial Symposium, Baltimore, Md., August.

U.S. Army Center for Health Promotion and Preventive Medicine/Oak Ridge National Laboratory. 1999. Derivation of Health-Based Environmental Screening Levels for Chemical Warfare Agents. A Technical Evaluation. (Principal author: D. Opresko; Point of Contact and Project Manager: Ms. V. Hauschild.) Aberdeen Proving Ground, Md.: U.S. Army Center for Health Promotion and Preventive Medicine.

Appendix A

Statement of Task

NHSRC SAFE BUILDINGS PROGRAM

Over the next three years, the EPA National Homeland Security Research Center's (NHSRC) Safe Buildings Program will provide research and technical assistance on methods, technologies, equipment and tools needed to protect built structures from threats and deliberate attacks of biological warfare agents, chemical warfare agents, and toxic industrial chemicals.

The Safe Buildings Program has created a Research Implementation Plan aimed at investigating containment, analysis, and decontamination issues relating to chemical and biological warfare agents and toxic industrial chemicals.

The Research Implementation Plan is designed to consider the universe of building protection and decontamination needs (as best understood at the present time), identify the highest priority needs for immediate action, and how best to accomplish those needs. This likely will mean that numerous high priority needs are best addressed by another federal agency, the private industry, or the emergency response community and this organization will be instrumental in meeting these needs. Each task will be addressed in a short (20 page) report discussing the issues outlined below.

The charge of the committee will be to review and comment in response to the following questions:

- From the expert committee's viewpoint, has the Research Implementation Plan completely and accurately identified important issues and needs in the buildings security arena? If the answer is no, what issues and needs should be added?
- From the expert committee's viewpoint, are the needs accurately prioritized and sequenced within the issues? If the answer is no, what would be recommended as adjustments and why is this the case?
- From the expert committee's viewpoint, are the projects recommended for funding over the next three years under each need appropriately sequenced to move to the ultimate product or products identified in the Appendix? If the answer is no, what would the committee recommend as adjustments and why is this the case?
- Overall, from the expert committee's viewpoint, what changes to the RIP would be recommended to improve its presentation in terms of content and structure so as to more clearly convey the buildings security research and technical support program that is described?

Appendix B

Committee Biographies

Steven R. Tannenbaum, Chair, is a professor of toxicology and chemistry at the Massachusetts Institute of Technology. He received his bachelor's degree and Ph.D. degree from the Massachusetts Institute of Technology. His research interests are in toxicology and chemical carcinogenesis. His laboratory addresses questions that relate to the role of both environmental and endogenous factors on human health. Dr. Tannenbaum is a member of the Institute of Medicine.

William Bahnfleth is a professor of architectural engineering at Pennsylvania State University and director of its indoor environment program. He received his bachelor's, master's, and Ph.D. degrees in mechanical engineering at the University of Illinois at Urbana-Champaign. His research concerns thermal storage, district heating and cooling, indoor air quality, building energy modeling and analysis, and application of thermal sciences to buildings.

R. John Collier is a professor of microbiology and molecular genetics at Harvard University. His research is currently focusing on diphtheria and anthrax toxins, applying genetic, biochemical, and biophysical methods to generate detailed models of each step in toxin action. He received the Eli Lilly Award in Microbiology and Immunology in 1972, the Paul Ehrlich Prize in 1990, and the Selman Waksman Award in 1999. Dr. Collier is a member of the National Academy of Sciences.

David A. Edwards is a professor of biomedical engineering at Harvard University. He obtained his Ph.D. in chemical engineering in 1987 and has since taught and performed research at the Technion (Israel), MIT, Penn State, and Harvard. His research is in aerosol drug delivery systems. He is the coauthor of two textbooks on advanced transport analysis and numerous scientific papers. He is a member of the National Academy of Engineering and the American Institute of Medical and Biological Engineering.

Leon R. Glicksman is a professor of mechanical engineering at the Massachusetts Institute of Technology. He received his bachelor's and Ph.D. degrees in engineering from MIT and a master's degree from Stanford University. His research interests are fluidized beds, heat transfer, energy efficient buildings, sustainable buildings, air circulation in buildings, and sustainable buildings for developing countries.

Lynne Haber is the manager of the research program at Toxicology Excellence for Risk Assessment in Cincinnati, Ohio. She received her Ph.D. from MIT. She has developed more than 20 noncancer and cancer risk assessments for EPA's Integrated Risk Information System (IRIS), and for other government agencies. Dr. Haber's current interest is in the use of mechanistic data in risk assessment and methods for extending dose-response curves to low doses. She has also done risk assessments using benchmark modeling and categorical regression, and published papers on the use of those approaches in risk assessment. Dr. Haber is also a member of the NRC panel on low-level exposures to chemical warfare agents.

Sangtae Kim is the vice-president and information officer of Lilly Research Laboratories, a division of Eli Lilly and Company. The departments reporting to Dr. Kim are responsible for the many facets of information technology in pharmaceutical R&D, including information technologies that drive discovery research, preclinical development, clinical research, regulatory affairs, and product teams. He joined Lilly in 2000 from a similar position at Parke-Davis Pharmaceutical Research, a division of Warner-Lambert Company. From 1983 to 1997 Dr. Kim was a faculty member in the Department of Chemical Engineering at the University of Wisconsin-Madison, including a term as department chair. His research interests include fluid mechanics, rheology, suspensions, protein dynamics, applied mathematics, and parallel computing. Dr. Kim is a member of the National Academy of Engineering and a past member of the Board on Chemical Sciences and Technology.

Charles E. Kolb is the president of Aerodyne Research, Inc. His research interests include the chemistry and physics of trace atmospheric species and the chemical kinetics and spectroscopy of combustion. Dr. Kolb has served on a variety of NRC committees, some of which deal directly with chemical and biological agents.

Ellen Raber is the department head for environmental protection at the Lawrence Livermore National Laboratory. In this role she is responsible for both operational and applied R&D efforts in pollution prevention, waste management, environmental restoration and environmental monitoring and analysis. Her research area has been in applied geochemistry, and she has authored numerous publications in this area. She has been actively involved in environmental applications for international CBW treaty verification and nuclear safeguards to include chemical and biological warfare agent detection and decontamination and related environmental issues.

William S. Rees, Jr., is a professor of chemistry and biochemistry at the Georgia Institute of Technology. He received his bachelor's degree from Texas Tech University and his Ph.D. from the University of California at Los Angeles. Professor Rees's research interests are in the synthesis and characterization of inorganic and organometallic compounds for use in the preparation of electronic materials. He has served on a number of committees dealing with issues of national defense.

Richard G. Sextro is a physicist in the Indoor Environmental Program at Lawrence Berkeley National Laboratory. He received his bachelor's degree from Carnegie Institute of Technology (now Carnegie Mellon University) and his Ph.D. from the University of California at Berkeley. Dr. Sextro has been actively involved in research concerning biological and chemical warfare agents in indoor environments. He has recently completed a modeling study on indoor dispersion patterns of anthrax spores.

Kent J. Voorhees is a professor of chemistry and geochemistry at the Colorado School of Mines. His research has a strong analytical chemistry component. In addition to polymer studies he has developed research programs in mass spectrometry, pyrolysis-mass spectrometry, supercritical fluid chromatography, chemometrics, and artificial intelligence. He also designed the biodetection sensors for biological warfare agents in use at the Pentagon.